STREAMER'S BIBLE ELGATO STREAM DECK

A complete guide for beginners

Alex Barry

Contents

Introduction

In the wake of the recent outbreak of deadly corona virus, working at home became necessary in no both private and public sectors across the globe. Working from home became a compulsory safety measure for all and sundry.

A big question that comes to our mind is how different people now work from home and achieve their goals and objectives? Let's take a look at a gadget that can be very helpful when it comes to working from home and its importance.

THE STREAM DECK

Managing video streaming is made easier with the stream deck. Buttons can be customized to functions as quick scene switch, inserting media, audio adjustment, and talking with viewer and a whole lot more. The stream deck can work efficiently with several controls and interacting with different activities.

You can assign commands you frequently use to a series of dedicated physical keys on the deck.

What displays on the "screen" of each button is completely your choice and the command you run frequently can be assigned or customized to a button.

The importance of the stream deck to consistent streamers cannot be really over emphasized.

The stream deck from Elgato is a control gadget with a physical customizable 15 buttons with different function assigned to it and also serves multi-tasking purposes.

It is essential for users who stream gaming sessions and also enable multiple controls

You can use the Elgato Stream Deck to create a script for one or more actions. The device's keys can be assigned to the scripts. When you press the key, a set of steps or actions will begin to play.

The free Stream Deck tool for Windows and Mac may be customized, and it syncs with popular streaming apps like Open Broadcasting Software (OBS) and XSplit to let you conduct and broadcast without touching your mouse or keyboard.

You might have dedicated buttons for going live, enabling different gameplay and camera scenarios, and turning on your "be right back" screen when it's time to take a break, for example.

This allows you to seamlessly go from talking to your viewers in a close-up camera shot to showing off some "Halo" action in a gameplay scenario, all without letting your viewers watch you clumsily mouse around in your streaming program.

The Elgato Stream Deck is a macro pad, to put it simply. You may program activities to happen when you press the device's keys using the software. These acts might be as basic as starting a program or as complex as launching a program. A macro pad's main goal is to save time and boost productivity.

START WITH STREAMING

As the name implies, It is a stream deck is a multi-tasking Gadget which runs on OBS supported by windows 8, 10. The OBS carries a web socket and also supports Golang library which handles events through assigned buttons and scenes are easily changed with an indicator showing active scene or sessions.

Screen sharing is a whole lot easier and faster because you don't need to pay attention on any other window while working on something else.

The stream deck developed by Elgato is a mini gadget with customizable keys which among other things can also be utilized to stream live contents by gamers.

It can also be helpful processing activities, reduced the rate of mouse clicking and also assigned keyboard shortcuts for easy and speedy workflow.

There are numerous built-in and downloadable icon sets for typical operations such as going live, hitting record, activating

music, and sending tweets, including some cool animated ones. You can also add your own icons to the mix.

The Stream Deck connects your tools and assists you in automatically detecting scenes, media, and audio sources so you can control them with a single keystroke. You can program 15 LCD keys to execute a range of activities, and you may use one touch, tactile operation to switch scenes, play media, adjust audio, and more. To make streams more personal, GIFs, photographs, movies, audio samples, and opening and outro graphics can be added. The Stream Deck is compatible with macOS 10.11 and later.

Main features:

Fifteen keys which are customizable, with thier own LCD screen.

118 mm x 84 mm x 21 mm / 4.6 x 3.3 x 0.8 in

190 g weight

15 x LCD keys that can be customized

USB 2.0 connector

Compatible with the following operating systems: Windows 10 (64-bit); macOS 10.11 or newer

Different actions can be assigned to each key. One unique feature of the deck is its LCD screen covered with a concave window. You might recall that in 2007, The Optimus

Maximus was launched with and OLED keypad. The stream deck draws it ideas from there.

The stream deck body built with plastic, While its fascia is made of aluminum

There is a vertical groove on the front to position your deck so that it projects towards you. There is a kickstand and rows of notches for sloting and adjusting angles on the rear. The legs of the kickstand fits into other notches and allowing you lock-in your preferred angle.

It comes with a USB for captive connection and a plastic lightweight stand with rubber feet to give it a firm grip on your desk when you a button are push.

Setup

Ordinarily, When you purchase the stream deck is perform no activity, The first activation process is to download the stream deck software from the elgato website which enable you to assign each key and pre-loaded hotkeys for major streaming softwares like YouTube, Twitch, etc.

Nevertheless, the stream deck can also be powered by different tools like he Premiere Pro, Lightroom or Final Cut Pro X. These setup processes are complicated since there are no pre-loaded shortcuts for in these applications.

But the Davinci Resolve profile can also run in the Elgato software with auto-load when Resolve is run. This applies to any software being used and create multiple shortcuts for a particular program.

Common keys can be configured with elgato software, some might be duplicates already in the keyboard but it becomes a larger and dedicated key sometimes with a logo of the particular program. Which also implies that a single button can now be used as an alternative for actions that are three keys combination like the 'freeze frame' action?

Navigating pages with a single button press brings satisfaction and ease.

It is for you?

One major importance of the stream deck is the fact that it is a good tool for online gaming, photo and video editing.

Customizable controls

The above image shows keys with image which can be created with the photoshop or any image editor software. You just need to create a 288 x 288 pixel PNG file for the buttons and assign actions to them easy and quick access to buttons that could sometimes be very important and professional, just like a button that quickly disables your webcam or microphone when there is a distraction either by a family member who entered the room or you need to eat something quickly. Better still,

A deafen and mute button can be set to silent background noise and makes you talk with your audience without any form of distraction and interruption.

It's simple to customize the Stream deck. Simply drag & drop actions onto keys to get started. To create a folder, drag and hold one key over another. Better yet, assign custom icons to your keys to make them uniquely yours.

There is a hotkey option that will help you save time. It can be assigned to a group of buttons. Ctrl+F, for example, or anything else that comes to mind. That combination will be activated every time you click on that exact button. When you're listening to music or audio files, the multimedia button on the system tab comes in handy.

You can pause/play tracks, skip to the next/previous song, or simply mute them. You can make your own folder in which to save whatever you desire.

Shortcuts and switching modes buttons, scenes change, instant message reply Or evening a live chat with your audience can be done with this deck. Generally, the alt+tabbing is required in widows for setting hotkeys. Other useful controls include buttons to change between scenes and game switch. These actions are seamlessly done with the stream deck.

2. Easy audience engagement

You can reply your audience with arranged messages in Twitch with a single press at any point in time. For example, People are constantly asking about tips and tricks on any particular game. An assigned button can do just that by just a single press.

Streams can be quickly clipped, run an advert, current chat deleted, drop a highlight maker for further editing, chat mode can also be to follower only, slow chat, or emote only.

All these easy and quick access controls makes it more professional to manage your stream without tension.

3. Folders within folders

The stream deck comes in different sizes, but the standard size of the stream deck has 15 buttons. While the mini has six buttons, the XL has 32 buttons. Folders can be created and command assigned to the same buttons, Folders within folders can be created.

Things are group logically, Twitch commands folder, for example can carry a subfolder full of gifs which can be displayed on the stream through the Streamlabs OBS. Stickers and gif could be created to react to different moods with a just a keypad push.

Smart lights home controls (RBG)

One amazing thing about the stream deck is powering home lighting. However, its compatibility is Nanoleaf and Philips Hue and the Elgato's Key Lights.

Meaning the deck can be programmed to give easy control of various lighting point in your home. The set up can be done with the standard software and voice assistant to turn on/off the lights or even to control the brightness of the light, though not in a satisfying way. The Elgato Key Light Air may be controlled with your smartphone as well as your PC or Mac.

The Key Light Air is operated by WiFi after connecting to it with your smartphone and pointing it in the direction of your wifi network. You'll be able to operate the devices individually from your iOS or Android device, as well as your PC or Mac's desktop, once they're connected.

The color temperature may be changed from 2,900K to 7,000K, with a maximum output of 1,400 lumens. This is about 1,000 lumens less than typical LED lights built for video production, but the Key Light Air isn't meant to light up big regions.

Video editing

Video editing can be done with elgato stream deck using different video editing software like the Ligthroom, Non-linear, editor, Premiere Pro and a whole more depending on your choice.

Controls can be added on any video editing command, Trimming anywhere in a video clip, spliting video tracks, multi-camera editing. Though these actions can be done with the keyboard, The Stream Deck makes it more efficient in your video editing via dedicated keys.

Multi actions enables series of different actions can be customized just by a single click, one button for example, could open audio software, video editing software, launch a project and more possibilities.

Considering all these functions, the stream deck is worth having irrespective of your field. This stream deck though not an essentiality for everyone majorly because of its price, It is a handy tool for consistent streamers that create content for their audience.

There are several options for capturing game footage; most especially the Elgato's Game Capture Software which uses a capture card (Game Capture HD60 S+) has very high pixels of 720p and other higher resolutions

One good thing about the game capture is its easy connection of its HDMI PORT IN to your console's HDMI PORT IN and its PORT OUT to your TV or monitor's PORT IN.

The Nvidia Shadowplay is a game recording software product by Macro Recorder; Though not a preloaded software in the stream deck it can be installed via a menu on the bottom right corner with "more actions button" for adding plugins.

Running the stream deck as an Administrator

Admin mode in every system is most time secured and grants some special access than a normal user mode. This case is not different with the elgato stream deck. Some software like the Visual Studio 2019, Photo and audio editing softwares will require administrator mode. Let's discuss some tips for running administrator mode.

When you run any software in admin mode, make sure you understand and accept the privacy and security risks that come with it. needs you to be aware of and accept the security and privacy risks that come with this option.

Please don't proceed if you don't understand what utilizing administrator mode entails for your system's security. You might not be able to complete the setup if you don't have permission to update your system's security settings.

Tweaking is required when you want to use the stream deck to control softwares using the administrator access in windows 10. The reason being that the streamdeck.exe also works with the administrator access and privileges.

The following steps run the stream deck service as an administrator. A prior knowledge of Windows operating system is an added advantage as we take a step by step analysis of system application like Scheduler and Regex.

Create scheduler task

Creating a scheduler task is first step since we can't get admin access in the user interface or the the start menu. The scheduler gives more options which is set to start the stream Deck on "executable" every time on startup.

Firstly, Type in "task scheduler" in the start search bar, open it and choose create below Actions. Below the general tab, give the task a name, STREAM DECK. As seen in the image below, check Run with highest privileges and select Windows 10 in the Dropbox.

Create task screen

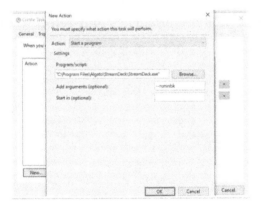

After clicking the action tab, click on new button, choose start a program in the action Dropbox and thereafter browse for the Stream Desk executable. Enter --*runinbk* in the add arguments box as displayed in the image below.

Action tab task scheduler

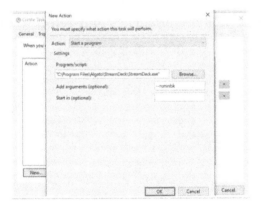

Remember to check the " allow task to run on demand" under the Settings tab. **PUSH OK**, and your task is created.

Since we now have task setup activated, we should know how to run it every time your PC comes up on log-on. The fastest and simplest method to do this is creating a short-cut in your desktop background and moves it to the startup folder.

Rights-click on the desktop background and click new > Shortcut. When the shortcut displays type in "schtask/run/TN Stream Deck" as showed in the image below. The task created is stream deck, hence give the short-cut the same name as stream deck and click finish.

Create shortcut

The last thing to do is to move the short cut to your start- up folder, To find your start-up folder, Click Windows-R and input shell:startup.

It will automatically bring back your startup folder. Take for instance;c:\users\yourusername\AppData\Roaming\Microsoft\Windows\Start Menu\Programs\Startup. The last step is to move the shortcut to your startup folder.

To locate your start-up folder, click Windows-R and type shell:startup. It will return your startup folder. For instance: c:\users\yourusername\AppData\Roaming\Microsoft\Windows\Start Menu\ Programs\Startup.

The last step is removing the default start-up that was initially created by the Elgato installer package.

There are different methods to do this;

Disable in task manager,

Input "Task Manager" in the start search bar, choose the Startup tab and find the stream deck entry and click disable

Remove the entry from the registry.

Unlike the task manager method, The registry editor is a permanent action. Input "Registry editor in the Start search bar, Find the StreamDeck.exe startup entry which is in: \HKEY_LOCAL_MACHINE\SOFTWARE\Microsoft\Windows\CurrentVersion\Run; click on the stream deck file and delete it.

As shown in the image below

The (BarRaider) plugin is an advanced launcher which allows you start programs as an administrator from your Stream Deck Startup programs.

Advanced Launcher (BarRaider)

If it is open, close the Stream Deck Client (configuration program).

In the Windows system tray, stop the current copy of the Stream Deck Windows service (see below) (right side of the task bar).

Right-click the Stream Deck Client icon or shortcut. Select "run as Administrator"

USING THE STREAM DECK FOR DAY TRADING (STOCK TRADING)

The Elgato Stream Deck is designed for streamers on YouTube and Twitch, as the name suggests. Don't let the name confuse you, All types of switching and opening techniques are possible with this gadget. If you've ever watched any of the finest streamers or content makers, you've probably noticed that they're utilizing some sort of device.

The stream deck is also suitable for everyday tasks. Programs, webpages, and a lot more can be opened. Trading stocks with it is quite impressive. Trading stock is done seamlessly; Each button can be programmed to do a specific action. Each of the buttons has its own set of instructions.

The device may be used to rapidly purchase and cancel open orders. As the stock rises, you can scale out of your positions and try to gain the larger move. As it progresses, you can send orders in various quantities.

The execution is simple because a cancel button has been built and allocated for that reason, which instantly cancels any open orders. If you truly need to get out of the stock quickly, a close button can be made.

A BUY button can be constructed to purchase any number of shares. The button, for example, was designed to trade 100 shares; it can be pressed twice to acquire 200, and so on.

OVERVIEW

It is advisable to be careful of this device at all times when you are on the stock platform. Pushing any keys activates the assigned actions(s). Entering a trade and exiting same is just key press. The stream is an important tool for those who really know its worth.

Lightning Source UK Ltd.
Milton Keynes UK
UKHW021316061222
413487UK00015B/234